Fundamental Structure of the Universe

Wasiful Alam

Alpona Publication

Contents

1 **A relation between gravity and magnetism**   7
   1.1 Introduction 7
   1.2 Main Hypothesis 8
       1.2.1 Rules Between the Relationships 8
   1.3 Methodology 9
       1.3.1 Explanation Through Equations 9
       1.3.2 Fragmented Behavior 13
       1.3.3 Effects of Gravitational field on motion 14
       1.3.4 Spiral Spin 14
       1.3.5 Autogenous Gravitation . . 15
       1.3.6 Into the Core 16
       1.3.7 Energy in different state . . 16
   1.4 Applications 17
       1.4.1 Process of changing field energy 19
       1.4.2 Cosmological Evidence in the very early universe 21
       1.4.3 Some possibilities 23

1.5 Conclusion 24

2 Theory of Universal Proportionality 25
2.1 Introduction 25
    2.1.1 Main Hypothesis 26
2.2 Methodology 26
    2.2.1 Construction of Relation . . 27
2.3 Energy in Newton's Law 27
2.4 Charged particle in Euler-Lagrange and Hamiltonian 28
2.5 Concentration of Field in Electromagnetism 29
2.6 Evidence in Charles's law 29
2.7 Force of Photons 30
2.8 Construction of the Proof 30
2.9 Conclusion 34

3 Creation of Mass 37
3.1 Numerical Calculation 37
3.2 Dimensional analysis of the equations 42
3.3 Discussion through Analogies . . . 45

4 Motion of Virtual Particles 47
4.1 Mass velocity exchange equation 48
4.2 In Relativistic Frame 48
4.3 Velocity of spiral Galaxies 49
4.4 In Non-relativistic or Virtual to Classical frame 50
4.5 For Wave Propagation 50
4.6 Mach's Principle 51
4.7 Based on the discussion 51

5 Fundamental Structure of the Universe 55
5.1 The third hypothesis 55
5.2 A prime function 55
5.3 Law of Conservation of Fundamental Structure 56
5.4 Co-relativity 56
5.5 Microstructures Hidden in all Basis 57
5.6 Three as the universal constant . . 58
5.7 Geometric perspective 59
5.8 Physical Representation 60
5.9 Historical Aspect 62

6 Final Value Approximation Method 65
6.1 HTML Structure 66
6.2 Initialization Using JavaScript 66
6.3 Modified Euler Method 67
6.4 Exact Solution 68
6.5 Plotting with Plotly 68

Chapter 1
A relation between gravity and magnetism

In different laws of force some physical quantities are influenced by fundamental changes, which are found to be common between these laws. Both falling objects and moving charges maintain acceleration. Matters react differently in different temperature. In equation relations are based on the increase or decrease of these values. As gravity and magnetism is dependent on these laws of force so a relation through the changing values are shown in this paper.

1.1 Introduction

Through Newtonian physics we understand how force and acceleration works on objects. Electric current, Spin magnetic moments of elementary particles, force between two magnets are caused in presence of magnetic field. Temperature is the manifestation of thermal energy, present in all matter, which is the source of the occurrence of heat, a flow of energy. From the equations which are developed, a relation can be made if the effects of change between these values are thought as proportional. Many experiments are needed to be done to get the appropriate values for proof. This paper focuses on deriving a theory from equations, also indicates where the proofs can be found, some new concepts and the outcomes of it.

1.2 Main Hypothesis

Neutral objects like neutrons and many other uncharged particles can have magnetic properties. Gravity and magnetism can occurs in an object at the same time. So there is a possibility of a relationship between these two fundamental forces of nature and magnetism could also be a cause of gravity. Through every experiment we see, flow of energy depends upon gravity and magnetism. Energy could be converted into something by a change in these force fields.

1.2.1 Rules Between the Relationships

- If two objects are neutral.

- Their gravitational and magnetic properties are same individually and for each other.

- The equation and result will be influenced by their flux, surface area, volume, concentration of mass, property of spin on their own axis, amount of mass and amount of energy they contain.

- Both of the objects have poles so their way of collision will also affect the value.

- But if the particles are charged then these changes will occur when two opposite particles collide or go through gravitational and magnetic field.

1.3 Methodology

The way equations of force between two objects or a charged particle are related is shown in this section. Some concepts through which we can understand the working mechanism of changing field energy are also given. Possible experiments are shown, which are based on the equations given.

1.3.1 Explanation Through Equations

From the law of gravity [15] we know, If force(F) between two objects with mass m_1 and m_2 is within a distance(d) then

$$F = G\frac{m_1 \times m_2}{d^2} \qquad (1.1)$$

Force between two magnets with surface area and flux density [23] [30] [24],

$$F = \frac{B^2 \times A}{2 \times \mu_0} \qquad (1.2)$$

where, A is the area of each surface. μ_0 is the permeability of space. B is the flux density.

According to the second rule if both forces are equal then from equation (1.1) and (1.2), using y as a parameter to relate the equations we get,

$$G\frac{m_1 \times m_2}{d^2} \propto \frac{B^2 \times A}{2 \times \mu_0} \times y \qquad (1.3)$$

$$m_1 \times m_2 \propto \frac{B^2 \times A \times d^2 \times y}{2 \times \mu_0 \times G} \qquad (1.4)$$

If Magnetic mass $M = m_1 \times m_2$ then,

$$M \propto \frac{B^2 \times A \times d^2 \times y}{2 \times \mu_0 \times G} \qquad (1.5)$$

These equations may vary according to the properties of those two objects.

After collision there must be a change in magnetic moment.

magnetic moment(m), [27] [41]

$$m = \frac{B_r \times V}{\mu_0} \qquad (1.6)$$

where, B_r is the residual flux density, V is the volume of the magnet. μ_0 is the permeability of vacuum.

From equation (1.4) and (1.6) we get,

$$m \propto \frac{V}{\mu_0 \times d} \times \sqrt{\frac{2 \times G \times m_1 \times m_2 \times \mu_0}{A \times y}} \quad (1.7)$$

But if either any magnetic or gravitational property changes then the force acting on the whole system will change. So, from equation (1.1) and (1.2), using x as a parameter to relate the equations we get,

$$F \propto \sqrt{\frac{G \times m_1 \times m_2}{d^2} \times \frac{B^2 \times A}{2 \times \mu_0}} \times x \quad (1.8)$$

Force due to gravitational acceleration,

$$F_G \propto m \times g \quad (1.9)$$

Force(F_M) due to the magnetic field (B) on a charged particle with mass(m) ,charge(q) and velocity(v),

$$F_M \propto q \times v \times B \times sin\theta \quad (1.10)$$

A charged particle with mass will also have gravitational acceleration [42] and if the velocity is caused by gravitational acceleration then from equation (1.9) and (1.10),

$$F_M \propto q \times \frac{F_G}{m} \times t \times B \times sin\theta \quad (1.11)$$

So, $F_M \propto F_G$ and $F_M \propto \frac{1}{m}$ in a charged object moving with an acceleration, this only occurs when one object's gravity is working on another object and the temperature of the total system is not changing. In this case, force of magnetism will increase if the object has less mass but change in gravitational field and flux is increasing.

But if gravity is acting upon the whole system rather than working between two objects and temperature of the whole system is changing, then,

From the equation of density,

$$mass\ density = \frac{mass}{Volume}$$

charge density is the quantity of charge per unit volume, [28] [36] [39]

$$charge\ density = \frac{amount\ of\ charge}{Volume}$$

From equation (1.11) and the two equations of density, for the mass(m) if the temperature is increased then the density will decrease causing the magnetism to increase.

From another way, if temperature is increased in a constant volume then the density of charged particles will decrease, covering up more of the volume due to their motion. So,

$$F_M \propto \frac{1}{density\ of\ charge\ of\ the\ particle}$$

and gravity also depends upon density of mass or in other form density of energy,

$F_G \propto$ $density$ of the $mass$ of the $charged particle$

Force of gravity depends upon the concentration of gravitational field,

$F_G \propto$ $concentration$ of $gravitational$ $field$

Density changes with temperature. [19] [45] so,

$$F_G \propto \frac{1}{Temperature}$$

If the temperature of a matter is increased, then its density of charge outside the nucleus decreases. The electrons get more energy and transforms it into motion, The flow of electron increases and magnetism increases as well. It also happens in the surface of Nutars and semiconductors. So in this case,

$$F_M \propto Temperature$$

If relativistic mass changes depending upon the velocity of an object, then there is a relativistic gravitational change. If the relativistic mass is m_{rel}, relativist gravitational force F_{rel_G} and the acceleration of the object a then,

$$F_{rel_G} \propto m_{rel} \times a \tag{1.12}$$

1.3.2 Fragmented Behavior

No matter how many fragments we create from an object which has both magnetic and gravitational properties, each of the fragments will conserve these two forces. These forces act even in quantum level. The tinier the object gets, the closer the value of these forces come.

1.3.3 Effects of Gravitational field on motion

If an object is moving through the gravitational field with an increasing velocity.

Because of increasing velocity, the object is moving through more amount of gravitational field in less time. If we think gravitation as a flow of energy, then the amount of energy the object is receiving and giving back due to gravitation is increasing more than that of its rest position.

And, the gravitational flux is never going to be the same from up, down and any other direction in a moving object. So, it will not be increasing the same amount in all direction.

1.3.4 Spiral Spin

If an object is spinning in its own axis at the speed of light. Then it is producing its gravitational waves [2] spirally. That means the change in gravitation in a distant point is not the same for each moment.

Let's imagine from a point on the surface of an object gravitational waves are coming out and the object is spinning on its own axis from left to right. Now if we freeze the total process at the

16th second then the first wave has traveled for 16 seconds the next wave has traveled for less than 16 seconds, it is shifted a bit in the right direction and if we keep adding all the wave points from the left to right we will get a spiral curve. The change in flux and field along the curve represents a direction in the curve from left to right. It proofs that gravity does not work straightly from the center of one object to another object. And it also creates a rotational motion between those two objects.

1.3.5 Autogenous Gravitation

If there are two objects, one is more massive and dense than the other and changing gravitational field lines of both are working upon each other or if gravitational wave is coming from only one object and there is nothing else then a process occurs. [22]

If we divide the small object into two sides then there is a side facing the massive object, let's imagine it as the left side. Now if we look closer to the point from where we have divided it, some gravitational waves or change of gravitational field lines are being produced from the right side and by being pulled these are ending up on the left side. From here we have got a direction of gravitational waves or field lines, from right to left. If the small object is spinning, then there will be a continuous change in the direction of these gravitational waves or field lines. Even if the object is not spinning, the right side is pulling the left side. This force is generating a spin and according to the direction of its spin the object is actually pushing itself towards the massive object. So objects in the universe can do only one thing and it's pull. But the same pulling force can be turned into pushing force by keep adding objects, increasing the space they contain. We also notice something similar to this in a magnet. The spinning and orbiting of the nucleus of an atom or an electron cloud produces a magnetic field. Every particle needs an acceleration to react with magnetic field If there is acceleration then there has to be a gravitational change.

1.3.6 Into the Core

Flux of gravitational and magnetic field of an object will keep decreasing with a decrease in radius. That's why there is no gravitational or magnetic force at the center of an object. It points to only one conclusion that there is nothing at the center of an object. No matter how tinnier the object and how closer an observer gets still there is nothing at the center but dimensions.

1.3.7 Energy in different state

In three different ways energy reacts with an object. In the first state energy works along the structure of an object. In second state it works along the atoms and in third state energy reacts with the inside of an atom. Relation between Gravity and

Magnetism in a steady object, two moving objects and in a moving charged particle is not same. We may be able to relate this changes if we measure in a very tiny or large scale.

1.4 Applications

- According to the equations there will be a change in mass after the collision. Mass of the product will increase or decrease according to the reaction. When four hydrogen atoms collide to produce a helium atom, there is a change in mass [37] of about 0.028758u.

- Electrons and protons carry same amount of charge. When they collide to form a neutron and a neutrino, [35] the neutron gets an extra mass of $1.3945616 \times 10^{-30} Kg$ which cannot be found in the additional mass of those two. Electron and proton add with each other because of magnetism while the process of change in mass happens because of gravity. There has to be a change in gravitational and magnetic properties, which can be seen in the product and reactant. The change in temperature and the other properties can be measured.

- Electromagnetic wave reacting with gravity [22] is another example of the relationship between magnetism and gravity. Light bends outside of the black hole because of gravity. In theory and calculation fields react and space bends, light is reacting with it only because gravity exists.

- Magnetic and gravitational changes also occur in atoms due to their energy levels of electrons. [6] If energy level changes then the radius from the core to the electron also changes, that's why the shape and density of the electron sphere changes within energy levels, orbits and orbitals. Magnetism and gravity works differently in different level and in different orbital.

- The mass of a proton is about 80–100 times greater than the sum of the rest masses of the quarks that make it up. Most of the mass is from the energy of field and motion. As mass is being grater than what it should then it's gravity is working in a different way

- In neutron stars and magnetars [38] [26] density of charged particles is less while in black holes density of mass is more than the density it had before. It is a proof of how gravity and magnetism shifts in each case.

- The concentration of magnetic field is higher in less massive objects or particles cause less massive objects have lower concentration of static energy, gravity or mass. Electrons are lower in mass. To find this out we have to keep one unpaired electron in every atom and measure their magnetic properties. The difference of magnetic moment between electron and proton is another example.

- If a spinning object is changing its direction over time, then for every 90-degree change there will be a change in its gravitational and magnetic force and field. Magnetic force will be rising up and going down and its acceleration [46] [32] will be changing to.

1.4.1 Process of changing field energy

Gravity gives acceleration to objects. Magnetism causes the electric force. That electric force converts heat energy into electric energy. Enough electric energy can produce photons. According to the equation $E = mc^2$, energy of photons can be turned into mass or vice versa. Electrons which have mass can be created from photons through the paired production process. [17] If protons and electrons collide they form neutrons which has higher mass then both of these. The change in heat energy is the cause of change in gravitational and magnetic field.

Gravity increases when the density [16] [34] of mass is increased(specific gravity) while magnetic force increases with momentum and decrease of the density of charge. If we examine Fe, Co, Ni and Gd, Fe has less charge density and less mass but more magnetic force. [40] Range of magnetism of these elements will vary with temperature. For magnetism the density of mass can be balanced by changing temperature. The magnetic properties increases with decreasing temperature and decreases if temperature is increased. But when a charged particle starts to change heat energy into kinetic energy then it's reaction with magnetic field changes too . In higher temperature the same object may produce magnetic field if the change in energy is happening in a constant volume and the extra heat in reaction is being emitted.

Inertia or in other words resistance in acceleration increases with mass so magnetism will decrease. In an atom if we keep decreasing the radius of an energy level then the concentration of charge will keep increasing and the energy of that atom will keep decreasing. If we convert this lost energy in mass then the gravitational and magnetic force we get from it will be as same as the force that is distributed in the motion of the system; the stationary energy is being changed into the energy it needs to keep its system in motion. In the sense of concentration of field and difference of temperature, Gravity is disproportionate to magnetism. Charged particles can gain more acceleration in higher temperature and also lose acceleration in lower temperature but density decreases with temperature. Objects can generate more gravitational force or concentrate more gravitational field by accelerating. If there are more than three dimensions then,

$$xG_a \propto \frac{1}{y\mu_{o_b}}$$ (using x and y as parameter, a and b as two different dimension). Absorption of

energy is different in different dimension.

If energy is conserved as mass then gravitation will increase but if the same energy is conserved as kinetic energy, reaction with magnetic field will increase. Because of acceleration gravitational flux will be more concentrated so gravity may not decrease in that case(Equivalence principle) [20] [21]. If stationery energy keeps increasing in a constant volume, then it's going to turn into a black hole but if kinetic energy keeps increasing in a constant volume then it's going to turn into a magnetar or something like that. The magnetic field around black hole is a strong proof of this. [7]

From energy in different state we can understand how turning a dynamo generates electricity and turns lights on. From the process of changing field energy, we can understand how gravitational flux and magnetic flux changes in it, the flow of energy in an object either as mass or as motion can be understood.

If gravity and magnetism both are related to temperature and velocity then these two are also related to each other. Through all of it we can see nothing more than an attraction force.

1.4.2 Cosmological Evidence in the very early universe

According to the standard model of cosmology [4] [1], during the earliest moments of cosmic time the universe was extremely hot and expanding very fast. It suggests that then gravity was interacting very weakly, which is shown in the relation of gravity and temperature.

In the electroweak epoch the electrostrong symmetry was broken and in the quark epoch hadrons were formed, from quarks massive amount of protons and neutrons were produced. It means at that higher temperature effect of magnetism was greater, which can be found in the relation between magnetism and temperature.

As the universe cooled, Electrons getting trapped in nucleus means magnetism of the whole system decreased. Gravity acting on the whole system increased at the end of the dark age when the temperature of the universe was very low, stars started to form.

In the given equations if temperature is turned to zero then force of gravity becomes infinite. That refers to the universe being very dense and compact before the big bang.

The whole concept suggests that before big bang some potential energy converted into kinetic energy in a constant volume or some kinetic energy outside the constant volume converted into potential energy by getting inside the volume , temperature started changing in a very little amount of time which ended up causing both big bang and the asymmetry between matter and antimatter [33] [31] [3].

Some of the described effects can also be examined at the surface of some nutron stars and black holes.

1.4.3 Some possibilities

- Gravity is the strongest force in the universe. Black holes produce force stronger than anything else, near the Schwarzschild radius no force is beyond gravitational.

- The weak and strong nuclear force could be caused by gravity and magnetism.

- Gravity could be the only fundamental force in this universe.

- If a black hole is combined with a magnetar and in the collision between two magnaters or black holes, the mass of the product would be different from the additional mass of the reactants.

- Magnetism in spinning electron cloud be influenced by Autogenous gravitation, near such potential field.

- In a very critical length and temperature, gravity and magnetism could be turned into each other. Actually not turned but their value will shift for conserving energy.

- A bit of light in search for dark matter.

- A possibility for the theory of everything

1.5 Conclusion

By examining different flow of energy in different states of neutral and charged particles or objects, the theory of magnetism and gravity being related is represented through some proportional equations. Change in magnetic moment and mass after collision between particles are shown in the example part as proof. The temperature difference between black hole and neutron star and the released energy in particle collision must be noticed. The concept of autogenous gravitation and spiral spin influences the flow of energy. The equations shown may not be equivalent but these are proportional. Changing a value in one equation changes the other values depending on it. Density, temperature and concentration of field are some great factors here.

Acknowledgements In the accomplishment of this paper, I would mostly thank my brother Samsuzzhan Alam Orko for teaching me all about the format of research papers otherwise it was not possible for me to publish this work.

I would also like to mention Wikipedia, because when ever I was looking for answers or information related to the research, I got that in there.

I was inspired reading about all those articles, books and research papers which I have mentioned in the reference section. Thanks to all those writers and publishers for the influence that I have got from their work.

Chapter 2

Theory of Universal Proportionality

The Universe is governed by changing forces acting over the fields. In classical gravitational, modern relativistic and Quantum physics we observe some similar behaviors while trying to measure the working force in a system or object. The nature of this Universe abides by some laws, explainable through a set of relations which is represented here.

2.1 Introduction

Quantum mechanics in subatomic level can influence large scale cosmic events. It explains the physical origins and evolution of the Universe and also can be observed in the more complete classical mechanics. This theory also proves some of the predictions of a previous paper "A Relation Between Gravity and Magnetism". Through change of temperature, energy exchange rate and concentration of field every event or system can be understood.

2.1.1 Main Hypothesis

Firstly through an observation we can understand how the concept of energy exchange rate, change in temperature and concentration of field is present in most of the previous equations related to force. Existence of these values are checked in previous equations. Then using static electric force from Coulomb's law in Bohr's model, max-plank's equation of vacuum permittivity , mass energy equivalence principle, specific heat related equations are plugged-in in both dynamic and static system. After rearranging some values the theory of universal proportionality can be clearly observed.

2.2 Methodology

In Newton's equation of force, multiple values for mass are changed, and the outcome is noticed. The effects in fields are understood through Lorentz force and relativistic equations for Euler-Lagrangian and Hamiltonian. The relation with temperature is shown using Charles's law and specific heat. The main proof is constructed based on Bohr and Coulomb's proposals.

2.2.1 Construction of Relation

According to the paper a relation between gravity and magnetism we can write

$$F_g \propto \frac{1}{\Delta T} \quad (2.1)$$

$$F_m \propto \Delta T \quad (2.2)$$

$$F \propto Concentration\ of Field \quad (2.3)$$

For energy exchange rate of the field we can write,

$$F \propto E_{er\uparrow\downarrow} \quad (2.4)$$

From the above relations,

$$F \propto \frac{1}{T} \times Concentration\ of Field \times E_{er\downarrow} \quad (2.5)$$

$$F \propto T \times Concentration\ of Field \times E_{er\uparrow} \quad (2.6)$$

Here the arrows represent the inward($\downarrow$) and outward($\uparrow$) directional values in the field.

2.3 Energy in Newton's Law

According to Newton's law of motion,

$$F \propto m \times a \quad (2.7)$$

If Einstein's equation of Specific Heat $Q = m \times s \times \Delta T$ is applied, where 's' is for specific heat of an object. [14]

$$F \propto \frac{Q}{s \times \Delta T} \times a \qquad (2.8)$$

Here we observe the relation $F \propto Q$ or $F \propto E_{er\downarrow}$ and also $F \propto \frac{1}{\Delta T}$

Using the short form of Mass–energy equivalence principle [12] we observe the same effect for energy exchange rate.

$$F \propto \frac{E}{c^2} \times a \qquad (2.9)$$

2.4 Charged particle in Euler-Lagrange and Hamiltonian

For a moving charged particle in an electromagnetic field, [44]

Hamiltonian

$$\mathbf{H}(t) = \dot{\mathbf{x}}.\mathbf{p} - L = c\sqrt{m^2c^2 + (\mathbf{p} - q\mathbf{A})^2} + q\varphi \qquad (2.10)$$

Euler–Lagrange equation

$$\dot{\mathbf{p}} = -\frac{\delta \mathbf{H}}{\delta \mathbf{x}} = q\dot{\mathbf{x}}.(\nabla \mathbf{A}) - q \nabla \varphi \qquad (2.11)$$

Here we observe that energy and field has a proportional relation with the action and the equations can be developed into the equation of electromagnetic field.

2.5 Concentration of Field in Electromagnetism

From Lorentz Force in electromagnetic field, [11]

$$F \propto q\mathbf{E} + q(\mathbf{V} \times \mathbf{B}) \quad (2.12)$$

or in scalar form,

$$F = q \times v \times B \times sin\theta \quad (2.13)$$

We observe $F \propto concentration\ of\ field$ through electric field E and magnetic field B acting over a charged particle, $F \propto \mathbf{E}$ and $F \propto \mathbf{B}$. When the fields increase in the same amount of area or volume the force increases as well.

2.6 Evidence in Charles's law

According to Charles's law [10]

$$V_\theta = V_o(1 + \frac{T_\theta}{T_1}) \quad (2.14)$$

Here we observe the situation for both ΔT and $\frac{1}{\Delta T}$.

Where if T_θ is -273 and T_1 is 273 the volume of an object becomes zero which can also be observed in $F \propto T$ scenario. The net Force acting over the fields in gas molecules become zero due to the temperature and that's why we get the volume zero.

2.7 Force of Photons

Pressure exerted by a photon [9]

$$P = \frac{Intensity}{Velocity\ of\ light} \qquad (2.15)$$

From here we get,

$$F = \frac{n \times \nu \times h}{A_{area\ of\ incident} \times t} \times A_{area\ of\ impact} \qquad (2.16)$$

or,

$$F = \frac{n \times E}{A_{area\ of\ incident} \times t} \times A_{area\ of\ impact} \qquad (2.17)$$

Here $F \propto \frac{E}{t}$ or $F \propto E_{er}$ and how energy is distributed in the field, in other words $F \propto Concentration\ of\ Field$ can also be observed in $F \propto \frac{A_{area\ of\ impact}}{A_{area\ of\ incident}} \times n$

2.8 Construction of the Proof

Static electric force in Bohr's model [8] [18] [25]

$$F_e \propto \frac{1}{4\pi\epsilon_0} \times \frac{e^2}{r_n^2} \qquad (2.18)$$

Charge of an electron orbiting an atom $e = \frac{\sqrt{4\pi\epsilon_0 m r_n}}{V_n}$

$$F \propto \frac{1}{4\pi\epsilon_0} \times \frac{4\pi\epsilon_0 m r_n}{V_n^2} \times \frac{1}{r_n^2} \qquad (2.19)$$

Velocity of the electron $V_n = \frac{nh}{2\pi m r_n}$

$$F \propto \frac{1}{4\pi\epsilon_0} \times 4\pi\epsilon_0 m r_n \times \frac{4\pi^2 m^2 r_n^2}{n^2 h^2} \times \frac{1}{r_n^2} \quad (2.20)$$

According to Maxwell's equation $\epsilon_0 = \frac{1}{\mu_0 c^2}$ [29]

$$F \propto \frac{1}{4\pi} \times \mu_0 c^2 \times 4\pi\epsilon_0 m r_n \times \frac{4\pi^2 m^2 r_n^2}{n^2 h^2} \times \frac{1}{r_n^2} \quad (2.21)$$

Applying the mass energy equivalence principle $E = mc^2$. In the equations 'subscript c' represents the mass and energy of moving objects

$$F \propto \frac{1}{4\pi} \times \mu_0 \frac{E_c}{m_c} \times 4\pi\epsilon_0 m r_n \times \frac{4\pi^2 m^2 r_n^2}{n^2 h^2} \times \frac{1}{r_n^2} \quad (2.22)$$

We can change the mass for specific heat related energy
$Q = mS\Delta T$ [13]

$$F \propto \frac{1}{4\pi} \times \mu_0 E_c \times \frac{S_c \Delta T_c}{Q_c} \times 4\pi\epsilon_0 r_n \times \quad (2.23)$$
$$\frac{Q_{rest}}{S_{rest}\Delta T_{rest}} \times \frac{4\pi^2 m^2 r_n^2}{n^2 h^2} \times \frac{1}{r_n^2}$$

Acceleration due to gravity, $g = \frac{Gm}{r^2}$

$$F \propto \frac{1}{4\pi} \times \mu_0 E_c \times \frac{S_c \Delta T_c}{Q_c} \times 4\pi\epsilon_0 r_n \times \quad (2.24)$$
$$\frac{Q_{rest}}{S_{rest}\Delta T_{rest}} \times \frac{4\pi^2 r_n^2}{n^2 h^2} \times \frac{g r_n^2}{G} \times m \times \frac{1}{r_n^2}$$

$$F \propto \frac{1}{4\pi} \times \mu_0 E_c \times \frac{S_c \Delta T_c}{Q_c} \times 4\pi\epsilon_0 r_n \times \quad (2.25)$$

$$\frac{Q_{rest}}{S_{rest}\Delta T_{rest}} \times \frac{4\pi^2 r_n^2}{n^2 h^2} \times \frac{r_n^2}{G} \times mg \times \frac{1}{r_n^2}$$

From the laws of motion, $g = \frac{v}{t}$

$$F \propto \frac{1}{4\pi} \times \mu_0 E_c \times \frac{S_c \Delta T_c}{Q_c} \times 4\pi\epsilon_0 r_n \times \quad (2.26)$$

$$\frac{Q_{rest}}{S_{rest}\Delta T_{rest}} \times \frac{4\pi^2 r_n^2}{n^2 h^2} \times \frac{r_n^2}{G} \times \frac{mv}{t} \times \frac{1}{r_n^2}$$

First order of development

$$[F \propto \frac{1}{4\pi} \times \mu_0 E_c \times \frac{S_c \Delta T_c}{Q_c} \times 4\pi\epsilon_0 r_n \times$$
$$(2.27)$$
$$\frac{Q_{rest}}{S_{rest}\Delta T_{rest}} \times \frac{4\pi^2 r_n^2}{n^2 h^2} \times \frac{r_n^2}{G} \times \frac{E}{C^2} \times \frac{v}{t} \times \frac{1}{r_n^2}]$$

Second order of development

For kinetic energy related change in the system, from [43] $E_k = \frac{1}{2}mv^2$ and $E_k = \frac{3}{2}K_B T_B$ we get $\frac{\frac{9}{3}K_B T_B}{v} = mv$

$$F \propto \frac{1}{4\pi} \times \mu_0 E_c \times \frac{S_c \Delta T_c}{Q_c} \times 4\pi\epsilon_0 r_n \times$$
$$(2.28)$$
$$\frac{Q_{rest}}{S_{rest}\Delta T_{rest}} \times \frac{4\pi^2 r_n^2}{n^2 h^2} \times \frac{r_n^2}{G} \times \frac{1}{t} \times \frac{\frac{9}{3}K_B T_B}{v} \times \frac{1}{r_n^2}$$

We can write it as,

$$F \propto \frac{9\mu_0\epsilon_0\pi K_B S_c}{n^2 h^2 G S_{rest}} \times E_c \times \frac{Q_{rest}}{Q_c} \times \quad (2.29)$$
$$\frac{\Delta T_c}{\Delta T_{rest}} \times T_B \times \frac{1}{t} \times \frac{4}{3}\pi r_n^3 \times \frac{1}{V}$$

From the above construction we can write,

$$Constant \sim \frac{9\mu_0\epsilon_0\pi K_B S_c}{n^2 h^2 G S_{rest}} \quad (2.30)$$

$$\Delta T or \frac{1}{\Delta T} \sim \frac{\Delta T_c}{\Delta T_{rest}} \times T_B \quad (2.31)$$

For the first order of development c^2 is a constant and the concentration of field increases with increasing velocity. It tells us the system it is working on has an inward energy exchange rate

$$Concentration\ of\ Field \sim \frac{4}{3}\pi r_n^3 \times V \quad (2.32)$$

For the second order of development concentration of field is depending on the volume the object occupies. So here we understand the energy exchange rate of the system/object inside that volume is outward

$$Concentration\ of\ Field \sim \frac{4}{3}\pi r_n^3 \times \frac{1}{V} \quad (2.33)$$

$$E_{er\uparrow\downarrow} \sim E_c \times \frac{Q_{rest}}{Q_c} \times \frac{1}{t} \quad (2.34)$$

2.9 Conclusion

The relations circle through all the previous equations given.

If F is counted as amount of processed data then according to the first relation, if we keep decreasing the temperature, increasing the concentration of field or in other words amount of transistors in the processor and also increase the energy exchange rate by feeding the system more power then data processing increases. Here if we turn the temperature to zero then the amount of data processing becomes infinite and as much we understand, black holes can process infinite amount of data.

If we use the concept of energy exchange rate, the energy exchange rate of a black hole is grater than the energy exchange rate of light in a constant volume that is why light bends.

In the construction of proof we observe multiple layers of energy. Through rest mass, changed mass of a moving object, mass affected by gravity, mass related to changed temperature and also from the energy of fields in empty space the concept of energy exchange rate is well understood. For concept of concentration of field if we keep increasing the velocity of an object, in the total volume occupied by it concentration of field increases or decreases depending on which system the calculation is based on. Use of Boltzmann's Constant is proper for a proton or a hydrogen ion. It represents the construction as, hydrogen ions spinning around a black hole. For different particles, this constant should be determined. The proof can be improved further by using vectors, actions and field related equations.

Acknowledgements In the accomplishment of this paper I would mostly thank my brother Samsuzzhan Alam Orko for encouraging me to find out the constant otherwise the construction wouldn't have been done. I was only showing how it works over the previous equations thinking there exists a constant.

I was inspired by reading the books I studied since sixth grade and all the equations I needed were present in my old eleventh and twelfth grade books. The provided articles, books and research papers mentioned in the reference section helped me to complete the citation. Thanks to all those writers and publishers for the influence that I have got from their work.

Chapter 3
Creation of Mass

Using the Feynman diagram, the Electron capture process changes a nuclear proton into a neutron and causes the emission of an electron neutrino.

$$p + e^- \to n + \nu_e$$

According to the process mentioned as magnetic mass,

$$m_p \times m_e \propto \frac{B^2 A d^2}{2\mu_0 G} y$$

We can write it as,

$$m_p \times m_e = \text{constant} \times y \times \frac{B^2 A d^2}{2\mu_0 G}$$

3.1 Numerical Calculation

First, we have to calculate the magnetic field for both electron and proton. Here velocity of proton is taken as 1, velocity of electron v_e is taken as the velocity in the first orbit of Hydrogen atom. r_e is the radius of the first orbit. $\sin\theta = 0.89$, $a_0 = r_e$.

$$B = B_p + B_e$$

$$B = \left(\frac{\mu_0 qv \sin\theta}{4\pi(r_e + 8.414 \times 10^{-16})^2}\right) + \left(\frac{\mu_0 e v_e}{2\pi r_e 2 r_e}\right)$$

$$A = 4\pi(8.414 \times 10^{-16})^2 + 4\pi a_0^2$$

$$d = a_0 + 8.414 \times 10^{-16}$$

To calculate only the value of y, we can write the equation as below and plug in all the values:

$$\frac{m_p \times m_e}{\frac{B^2 A d^2}{2\mu_0 G}} = y$$

We get the approximate value of $y = 1.6670953509933026194 \times 10^{-35}$, which is near Planck length. And we know that,

$$m_n + m_{\nu_e} = 1.67492747142 \times 10^{-27}$$

$$(m_n + m_{\nu_e}) - (m_p + m_e) = 1.39463485789 \times 10^{-30}$$

Which is the untraced mass, and there could be additional mass lost as energy x, or we are unable to calculate. We can write it as:

$$m_T + m_G + m_{\mu\epsilon} = 1.39470154 \times 10^{-30} + x$$

$$m_{\text{total}} = m_p + m_e + m_T + m_G + m_{\mu\epsilon} + x$$

From the analysis of the equation, we understand that m_{total} has the value of the total working electromagnetic field, gravitation, temperature-related exchanged energy, radiated waves, weak interaction, strong interaction, and with it the mass of the total product and reactant. So,

$$m_{\text{total}} = m_p + m_e + \frac{3}{2}\frac{k_B T_B}{C^2} + \frac{2E_k}{v^2} + \frac{E_{\text{potential}}}{v_e^2} +$$

$$\frac{h\nu}{C^2} + \frac{K_e e^2}{(r_e + r_p)2(r_e + r_p)\cos\theta}$$

The energy of two gamma rays emitted during the process is $E_\gamma = 0.136 MeV = 2.17896 \times 10^{-14} J$, E_k is the energy of the electron in the first orbit. The temperature is taken as the value closer to plasma $T_B = 6000$ Fahrenheit.

The value of g is:

$$g = \frac{G(m_p + m_e)}{(r_e + r_p)^2} = 2.11136361686 \times 10^{-27}$$

And m_G is:

$$m_G = \frac{(m_p + m_e)g(r_e + r_p)}{v_e^2}$$

which is used as potential energy. Add the energy related to the static electric force between two opposite charges, $\cos\theta$ is taken as 1.

Adding up all the acquired values we get:

$$\frac{3}{2}\frac{k_B T_B}{C^2} + \frac{2E_k}{v^2} + \frac{E_{\text{potential}}}{v_e^2} + \frac{E_\gamma}{C^2} +$$

$$\frac{K_e e^2}{(r_e + r_p)^2 (r_e + r_p) \cos\theta C^2}$$

$$= 1.38256079576 \times 10^{-36} + 9.0863826859 \times 10^{-31}$$
$$+ 2.12141974727 \times 10^{-80} + 2.42441996614$$
$$\times 10^{-31} + 3.86142943189 \times 10^{-29}$$
$$= 3.9765375967 \times 10^{-29}$$

(3.1)

$$m_{\text{total}} = 1.71329821253 \times 10^{-27}$$

if y is multiplied by the constant, then we should get the additional mass, which can be expressed as:

$$\text{Constant} = \frac{m_T + m_G + m_{\mu\epsilon}}{y}$$

$$\text{Constant} = \frac{1.39463485789 \times 10^{-30}}{1.6670953509933026194 \times 10^{-35}}$$

$$= 8.3656574116 \times 10^4$$

If we analyze the development until now, the power of constant is on the positive side of the decimal 10^4, which is directly related to Coulomb constant $K_e (8.987551792314 \times 10^9 \text{kg·m}^3\text{·s}^{-2}\text{·C}^{-2})$, conductance quantum $G_0 (7.748091729 \times 10^{-5} S)$, molar gas constant

$R = N_A K_B (8.324462618 J\text{mol}^{-1} K^{-1})$, as we can check the additional values of these in the total mass.

The Boltzmann constant is related to black body radiation, entropy. Planck's constant and charge of the electron are used in quantum conductance. Therefore, the constant can be written as:

$$\text{Constant}_{\text{mass}} = \frac{K_e G_0}{R}$$

If we multiply y with the constant,

$$y \times \text{Constant}_{\text{mass}} = 1.667095351 \times 10^{-35} \quad (3.2)$$
$$\times \frac{8.987551792314 \times 10^{-9} \times 7.748091729 \times 10^{-5}}{8.314462618}$$
$$= 1.3962475320 \times 10^{-30}$$

Here we have got:

$$(m_n + m_{\nu_e}) - (m_p + m_e) \leq y \times \text{Constant}_{\text{mass}}$$

From the equation of the magnetic moment:

$$m \propto \frac{V}{\mu_0 d} \sqrt{\frac{2 G m_1 m_2 \mu_0}{Ay \times \text{Constant}_{\text{mass}}}}$$

$$m_{\text{magnetic moment}} = \text{Constant}_{\text{magnetic moment}} \frac{V}{\mu_0 d} \times$$

$$\sqrt{\frac{2 G m_1 m_2 \mu_0}{A(y \times \text{Constant}_{\text{mass}})}}$$

Using 0.227 as the ratio of weak interaction according to the Weinberg angle and previous measurement of mass, we get the magnetic moment of the neutron.

$m_{\text{magnetic moment}}$

$$= 0.227 \times \frac{4}{3}\pi(52.90087$$
$$\times 10^{-12})^2 \frac{\mu_0}{\sqrt{2Gm_p m_e \mu_0}} 8791.46430011 \quad (3.3)$$
$$\times 10^{-24} \times 1.3962475320 \times 10^{-30}$$

$$= 9.6630901764 \times 10^{-27}$$

3.2 Dimensional analysis of the equations

If we do the dimensional analysis of y,

$$\frac{m_p \times m_e}{\frac{B^2 A d^2}{2\mu_0 G}} = y$$

$$y = m_p \times m_e \frac{2\mu_0 G}{B^2 A d^2}$$

$$[y] = M^2 \frac{ML^2T^{-2}I^{-2} \times M^{-1}L^3T^{-2}}{(MT^{-2}I^{-1})^2 L^2 L^2}$$

$$[y] = [L]$$

42

So, y is a length and from the calculation, we understand that it is closer to Planck length $l_p = 1.616255(18) \times 10^{-35}$. The equation relies on the length, where the event took place.

$$[y \times \text{Const}_{\text{mass}}] = L\frac{ML^3T^{-2}(TI)^{-2}M^{-1}L^{-2}T^3I^2}{J\text{mol}^{-1}K^{-1}}$$

It shows us what mass really is. Mass is quantization of energy through a frequency of changing area in space with the change of temperature. Simply, it is a changing ripple of space. Here, mol refers to the amount of quantized energy, and J is for energy being in process.

The term tells us that if in a system more energy is being quantized, its mass increases. More J means the system has less mass but produces more energy, like stars. Decreasing J means the system is producing more mass as more energy is quantized, which acts like a black hole, where infinite mass is accumulated in no time.

$$[y \times \text{Constant}_{\text{mass}}] = L^2 \times K \times \frac{1}{T} \times \frac{\text{mol}}{J}$$

If we go a little deeper to understand its relation with time: Temperature is divided in the ratio of quantized energy depending on the time spent. More time spent in the process means more mass is gained, as happens when objects reach a temperature to change their form.

$$[y \times \text{Const}_{\text{mass}}] = L^2 \times K \times \frac{1}{T} \times \frac{\text{mass}}{\text{molar mass} ML^2T^{-2}}$$

$$= \left[\frac{KT}{\text{molar mass}}\right]$$

General magnetic moment,

$$\left[\frac{\text{Joul}}{\text{Tesla}}\right] = \frac{ML^2T^{-2}}{MT^{-2}I^{-1}} = [L^2I]$$

Squared magnetic moment,

$$[m_{\text{magnetic moment}}]^2 = \left(\frac{L^3}{ML^2T^{-2}I^{-2}L}\right)^2 \times$$

$$\left(\frac{M^{-1}L^3T^{-2} \times M^2 \times ML^2T^{-2}I^{-2}}{ML^2}\right)$$

$$[m_{\text{magnetic moment}}]^2 = \left[\frac{L^3I^2}{M}\right]$$

When the derived definition of mass is used,

$$[m_{\text{magnetic moment}}]^2 = \frac{L^3I^2}{L^2KT^{-1}J^{-1}\text{mol}} = \left[\frac{LI^2JT}{K\,\text{mol}}\right]$$

$$[m_{\text{magnetic moment}}]^2 = \left[\frac{LI^2\,\text{molar mass}}{KT}\right]$$

In general, we get the relation between length and current, but here we observe how length and current are related to the ratio of quantization, temperature, and time.

3.3 Discussion through Analogies

If we use the derived form of mass to analyze energy, then:

$$[E] = [ML^2T^{-2}] = L^3 \times K \times \frac{1}{T} \times \frac{\text{mol}}{J} \frac{L}{T^2}$$

It represents the acceleration of the process, changing the position of a volume of space in frequency. If the cube has L length and moves an L amount of distance, tensors and vectors are needed to understand its position in space. Symmetry and rotation-related problems arise, like determining if the object has rotated, if the second phase is the same as the first phase, or if it's a mirror image. The direction of movement also needs resolving. One thing is certain: It always moves depending on its length, maintaining the frequency of the quantization process.

To understand the system further, Pascal's mechanism of explaining pressure can be used.

The height of the cylinders is the same as the distance between them. Length of decreased energy L_x and increased energy L_y are the same, $L_x = L_y$. Maintaining the conservation of energy, $L = L_x + L_y$. This tells us there exist certain bands or levels of energy, similar to the spectral series in atoms. It is possible to increase the distance between two cylinders, and to do so, the height can be increased, but the width must decrease.

As the cylinders are thinned, stretched, and the distance between them increases, a tiny amount of energy works over a very large distance. Here, probability comes into play. It represents the electron cloud, quantum jumps, and shifts of energy levels in semiconductors and transistors.

[Note: Working with the constant of mass first, I tried to develop it according to the Feynman diagram, tried it in different possible ways but at the end I had to divide the entire equation with 1.61 to get the proper value. The same number came in several ways of calculation, a day letter when I checked the number, it looked familiar, so I searched and golden ration ϕ turned out to be 1.618... ... . When ever I used ratios related to Weinberg angle, mass ratio I had to divide the outcome with 137.5 which again I found to be called golden angle, or I had to use the Weinberg ratio related to near 32-degree angle (depending on the velocity of particle it changes) and I saw that, 36 degree is the smallest angle of golden triangle. To me, it felt like the way tree leaves grow at 137.5 degree to maintain the most energy exchange rate or sunflower seeds maintain the golden ratio to maximize the capacity of seeds, the same way ϕ occurs during quantization. As I had no proper explanation so I left developing that way, taking it just as a coincidence. I worked on the previous papers trying to know How nature itself understands physics, If that was truly an outcome of it then we can tweak our tools to it's level for achieving higher accuracy]

Chapter 4

Motion of Virtual Particles

Virtual particles do no work. Due to the uncertainty of position and momentum, they are in an always changing state, even during interaction with other fields they are found to be in uncertainty. Even if any force acting on macroscopic confined space they do no work on the field. They do not precisely obey energy momentum relation. Due to these facts,

$$m_v^2 c^4 \neq E_v^2 - p_v^2 c^2$$

Unusual kinetic energy relation to velocity such as quantum tunneling in accelerating frame of reference, they represent paired production. As their position and momentum not always the same, so it can be explained in a more general way. Let's imagine an infinitesimally small virtual particle moving through an infinitesimally small cross-section of space. The virtual particles are continuously being absorbed and emitted by this cross-sections.

4.1 Mass velocity exchange equation

$$\alpha_1 m_1 v_1^{(n_1)} + \alpha_2 m_2 v_2^{(n_2)} + \ldots\ldots$$

$$= \sum \alpha_i(n) m_i(n,v) v_i^{n_i},$$

where $\alpha, n \in \mathbb{R}$

4.2 In Relativistic Frame

$$m_1 = \frac{m_0}{\sqrt{1-\frac{v_0}{c}}}, m_2 = \frac{m_1}{\sqrt{1-\frac{v_2}{c}}}, \ldots\ldots$$

$$m_j = \frac{m_{j-1}}{\sqrt{1-\frac{v_{j-1}^2}{c^2}}}$$

Thus, it can be written as,

$$\sum \alpha_i(n) \left(m_i + \sum \frac{M(m_i)_{j-1}}{\sqrt{1-\frac{V(v_i, m_i, m_{j-1})^2}{C^2}}} \right) v_i^{n_i}$$

here the complexity rises when the particle is in between the cross-section $\square$ at the

$$\nu_\tau \longrightarrow \square \longrightarrow \square \longrightarrow \square \longrightarrow \nu_\mu$$

transition state it has a change of mass due to continuous absorption and emission, and again it has an infinitesimally small change of $mass_v$ due to relativity, such as Neutrina mass oscillation.

Due to the law of conservation of energy and momentum for the virtual particle to ∃xist and for the energy of empty space or Field energy to conserve the second law of thermodynamics

$$U(S_1, x_1, x_2, \ldots, x_r) = \sum \frac{\partial U(\alpha S, \alpha x_1, \ldots, \alpha x_n)}{\partial(\alpha S)}$$

α is the fractionability of space or the field, depending on the behavior of particle.

4.3 Velocity of spiral Galaxies

The Tully-Fisher relation for the velocity of spiral galaxies suggests,

$$V \propto \sqrt[n]{L}$$

where L is luminosity and n is generally taken from 1 to 4. For spherical blackbody,

$$L = 4\pi R^2 \sigma T^4$$

R Radius of the object, σ Stefan-Boltzmann constant, T Surface temperature of the object in Kelvin.

4.4 In Non-relativistic or Virtual to Classical frame

Fields energy is less than rest energy thus

$$m \sum \alpha_i V_i^{n_i} = \frac{1}{2}mV^2, \text{where} \quad \alpha_i < \frac{1}{2} \text{ and } n_i < 2$$

this behavior of virtual state to Exist can be observed in double pendulum, it can also be considered

$$m(\mathcal{O}(n)) \sum \alpha_i(\mathcal{O}(n)) V_i^{n_i}(\mathcal{O}(m,\alpha))$$

As the properties of both the G field and the device is always changing or error in case of Simulation. For thermodynamics, it can also be,

$$m \sum \alpha_i V_i^{n_i} = \frac{3}{2}KT$$

4.5 For Wave Propagation

If we consider the volume as a wave guide, then it will consist of modes.

$$dw = \frac{c}{\sqrt{1 + \left(\frac{w_{\min}}{C_k}\right)^2}} dk, \quad w = \sqrt{1 + (w_{\min}C_k)^2} K$$

$$V_\phi > c, V_g < c$$

Where for phase velocity to be greater than speed of light it is extremely necessary for this relation to hold

$$\frac{1}{2}\hbar\omega_n = \sum \alpha_i m_i V_i^{n_i}$$

If the time interval of two notes is less than the trilling time in a brass tube then the superposition of the two notes will create a new third note due to the relation

$$\Delta \omega \times \Delta t \geq \pi$$

which is similar to the Neutrina example

4.6 Mach's Principle

In $E = I\omega^2$, inertia is yet to be understood.

$$I_{ij} = \sum_\alpha m_\alpha \left(\delta_{ij} \sum_k (X_{\alpha,k}^2 - X_{\alpha,i} X_{\alpha,j}) \right)$$

here I_{ij} is a proportionality constant between $E_{rotation}$ and angular velocity, $I = \sum mr^2$. From outside the rotating reference frame, centrifugal force is a fictitious force field. Non-rotating from inside, rotating from outside.

4.7 Based on the discussion

$$ds = \frac{\delta Q}{T}$$

$$S = K \ln(\Omega)$$

$$Q = \alpha kT \ln(\Omega)$$

$$\ln(\Omega) = (\Omega - 1) - \frac{1}{2}(\Omega - 1)^2 + \frac{1}{3}(\Omega - 1)^3 + \ldots$$

$$Q = \frac{1}{3}\alpha m V^2 \ln(\Omega)$$

$$Q = \alpha \frac{1}{3} m \times$$

$$\left((\Omega - 1)V_1^2 - \frac{1}{2}(\Omega - 1)^2 V_2^2 + \frac{1}{3}(\Omega - 1)^3 V_3^2 + \ldots\right)$$

Let's compare one state, the third state

$$\frac{1}{3}(\Omega - 1)^3 V_3^2 = \frac{1}{n} V^n (\Omega - 1)^n$$

Thus,

$$V^n = \frac{n}{3}\frac{(\Omega - 1)^3}{(\Omega - 1)^n} V_3^2$$

The virtual velocity V is given by:

$$V = \sqrt[n]{\frac{n}{3}\frac{(\Omega - 1)^3}{(\Omega - 1)^n} V_3^2}$$

The exchanged energy,

$$Q = \frac{1}{3}\sum \alpha_i m_i \sum \frac{n}{i}\frac{(\Omega - 1)^i}{(\Omega - 1)^n} V_i^2 (-1)^{i-1}$$

virtual velocity is discrete change of velocity due to future state of field elements, depending on the second law (α, n) are structural properties of the system and particle. Virtual displacement

$$\delta x_i = \frac{\partial x_i}{\partial q_i} \delta q$$

takes no time, as it is a discrete change of position. Same thing happens in magnetic dipole, monopole. Here, Einstein will argue god and black magic, but will not provide any further help to modern physics.

Math is extremely helpful in switching positions. But here we have the problem of Uncertainty. Velocity here is not how you have always been forced to imagine by the radical chair holders in physics institutes. Due to the problem of uncertainty power is swap-able as well as position with respect to the scalable velocity V^n. Here velocity can compensate for power n and the power can compensate for velocity V as it is in an uncertain state. V_3 is fixed velocity of the third state, V_3 can't change but n can change. $V_3^2 \cong V^n$

After power swap, they still remain the same due to scalability of n and V. Existence of a particle rely on Parallel velocity of the state.

$$V_3^n \cong V^2$$

Thus Parallel velocity for a given state,

$$\frac{1}{V_3^n} = \sum \frac{n}{3} \frac{(\Omega - 1)^3}{(\Omega - 1)^n V_i^2}$$

scalable velocity implies it changes and any change will indicate another sub state so we sum up for i

Chapter 5

Fundamental Structure of the Universe

5.1 The third hypothesis

For the universe to organize a given set of elements in a same length of box due to distribution of prime numbers, the box uncertainly overflows causing Infinite fractals.

5.2 A prime function

Accordingly, the elements can be organized in n! possible ways in a n^2 box for a constant C as a scaling factor. Thus, it can be written as prime function f(n),

$$(\forall n \in \mathbb{N}) \vee (n \in \mathbb{P}), (n > 2)(\exists f(n))$$

$$[\prod_p \left(\frac{1}{1-p^{-s}}\right) = \sum_n \frac{1}{n^s} = \zeta(s)]$$

$$\iff f(n) = \frac{n!}{\left(\frac{n}{3}\right)^2} : f(n) \in (\mathbb{R} \setminus \mathbb{N}) \vee f(n) \in \mathbb{N}.$$

Here n is a prime if and only if f(n) is a fraction, otherwise it is not.

5.3 Law of Conservation of Fundamental Structure

A fundamental structure always fully origin, even within any infinitesimaly small or large change to it. The fundamental structure remains through any type of transformation. The operator applied on a system can not change the fundamental structure. It keeps emerging over and over again, causing a system to obey all the laws of known universe.

5.4 Co-relativity

Using $\sim<>\sim$ as the co-relative symbol if

$$f = (x+y)^b, a = g(f) \times q^2 \implies a \sim<>\sim b$$

Where a and b are correlated in an unknown Hamiltonian cycle. Such an example would be,

$$\sum_{n=1}^{\infty} \frac{1}{n^s} \neq \frac{1}{P+1}$$

but due to the structural similarities

$$(\frac{1}{\Box}, \frac{1}{\Box + c}) \implies n \sim<>\sim P$$

though they are surely not the same, but they are structurally correlated to each other.

5.5 Microstructures Hidden in all Basis

The fundamental base generating function should contain the prime function as a microstructure, here we can see,

$$\ln(x) = \ln(a) +$$

$$\sum_{n=1}^{\infty} (-1)^{n+1} \frac{n!}{n^2} \sum_{k=0}^{n} (-1)^k \frac{a^k x^{n-k}}{(1 - \frac{k}{n})(n-k-1)!\, k!}$$

$$\implies \ln(x) \sim<>\sim \frac{n!}{n^2}$$

As any transformation should preserve the prime function, such transformation can be observed

$$\sec^{-1}(x) \neq 1 + \frac{2!}{x^2} + \frac{4!}{5x^4} + \frac{n!}{ax^n}$$

but as the correlation is observable,

$$\sec^{-1}(x) = \frac{\pi}{2} - \sum_{n=1}^{\infty}(-1)^n \frac{(2n)!}{n \cdot n^2 \cdot (n-1)!^2 \cdot 2^{2-1} \cdot x^{2n}}$$

$$\implies \sec^{-1}(x) \sim<>\sim \frac{n!}{n^2}$$

f(n) generates the whole space.

5.6 Three as the universal constant

At least three minimum parameters are required to write a logical expression (a, b, =) : (a=b).

The minimum parameter for inner connected topology or the smallest imaginable structure triangle ($\triangle$) contains three sides. Three special coordinates to explain a system, three colors (rgb) required to observe the spectrum.

For ($n > 2$), 3 components required to have an atom (e p n). (IiIiI) 3 quarks to form the two components (u d d) or (u u d).

These repeating pattern comes to the point transistors required at least three layers (p n p) or circuit Direction (emitter, base, collector) to operate. When transistors in the processing units are utilized by Al-gorithm or simulation, they heat up causing black body radiation

$$f(n) = \frac{3^2 \cdot n!}{n^2}$$

all the atomic quantities to make the transistor are measured in Moles, $\ln(n) \sim<>\sim (n-1)!$ and for Avogadro number N_A

$$\lim_{n \to N_A} \frac{3^2}{n} = 1.49 \times 10^{-23} \approx k_B$$

Thus,

$$f(n) = \frac{3^2}{n}(n-1)! \implies k_B \ln(\Omega)$$

Entropy in information theory or in the transistor due to the movement of electrons both depend on three, and any Al-gebraic or Al-gorithmic expression contains three as a limiting factor.

5.7 Geometric perspective

If n is considered the hypotenuse using the right triangle theorem it can be written as $n^2 = a^2 + b^2$ or it can be considered as the radius of a sphere, $n^2 = x^2 + y^2 + z^2$ To understand motion in space time the light cone is used, as, a flash of light creates a sphere. Here, $(cdt)^2 = x^2 + y^2 + z^2$ the Lorentz factor γ is derived by incorporating the right triangle theorem

$$(ct')^2 = ct^2 + (vt')^2 \implies t' = t\sqrt{\frac{1}{1-\frac{v^2}{c^2}}}$$

where

$$\gamma = \sqrt{\frac{1}{1-x^2}} \sim<>\sim \frac{1}{1-\square} \sim<>\sim \frac{1}{1-p^{-s}}$$

according to the geometric series

$$\int \frac{1}{1-x} dx = \int (1 + x + x^2 + x^3 + \cdots + x^n) \, dx$$

$$= -\ln(1-x) = x + \frac{x^2}{2} + \frac{x^3}{3} + \cdots$$

$$\implies \gamma \sim<>\sim f(n) \sim<>\sim ln(\Box)$$

Gaussian curvature in Bertrand theorem

$$k(p) = \lim_{r \to 0^+} 3 \frac{2\pi r - c \cdot r}{\pi r^3}$$

is the deviation of the surface from flatness at a point. Which is studied in Riemannian manifold, curvature of the space of Qbits. The expansion of volume of the geodesic ball

$$v_r(0) = 4\pi \int_0^R \frac{t^2}{(1+t)\sqrt{1-t^2} f \frac{(1-t)}{(1+t)}} dt$$

also incorporates the multiplicity of 3.
$k(p) \sim<>\sim V_r \sim<>\sim f(n)$ the structure of geometric series is a binding condition.

5.8 Physical Representation

In general relativity, the structure is more visual through Flamm's paraboloid

$$-dS^2 = \frac{dr^2}{1 - \frac{r_s}{r}} + r^2 d\phi^2, \quad r_s > r$$

$$\sqrt{\frac{1}{1-\frac{r_s}{r}}} \sim<>\sim \frac{1}{1-p^{-s}} \sim<>\sim f(n)$$

from the perspective of linear Al-zebra, dimensions, analytical geometry n! can be written in many possible ways,

$$n! = \sum_{\sigma \in S_n} \text{sgn}(\sigma) \prod_{i=1}^{n} a_{i,\sigma(i)} = \prod_{k=1}^{n} k = \Gamma(n+1)$$

$$= \int_0^\infty x^n e^{-x}\, dx$$

The function $\text{sgn}(\sigma)$ returns $+1$ for paired permutations and -1 for unpaired permutations. This property captures the orientation of the arrangement of elements.

These expressions are required to understand motion of objects in a space.

In statistical quantum mechanics, it can be represented as

$$n! = \frac{1}{h^{3N}} d\tau_N \prod_i d\mathbf{p}_i\, d\mathbf{q}_i \approx \left(\frac{n}{e}\right)^n$$

Partition function resembles the same structure

$$Z(\beta) = \prod_i \frac{1}{1-e^{-(\beta\epsilon_i+\alpha)}}$$

$$Z(\beta) \sim<>\sim \frac{1}{1-p^{-s}} \sim<>\sim f(n)$$

Regardless of the scale, all interactions are explained using

$$\frac{d}{d\theta}\ln(\cos\theta) = \tan\theta \implies f(n)$$

This co-relativity is observed in any ratio. Weight $\sqrt{\frac{1}{1-x^2}}$ of Chevy chef polynomial, $\sum_{n=0}^{\infty} a_n t_n(x) = \log(1+x)$ is correlated to Ramanujan partition function,

$$\prod_{k=1}^{\infty} \frac{1}{1-x^k} = 1 + x + 2x^2 + 3x^3 + \cdots = \sum_{n=0}^{\infty} p(n)x^n,$$

$$\frac{1}{1-x} = \sum_{n=0}^{\infty} x^n$$

which is further used as approximation method in physical interactions. In all branches of science, these structural similarities emerge over and over again

5.9 Historical Aspect

Near a cosmic river lived the creators of the number system, who spoke in the language of Mathematics. From a rational perspective, it should sound mythological. Both the word Mathematics and Number are fairly new. From the creators' perspective it must have been more artistic ,physical, structural. They included the letter 0, void in order to generate the basis. To them, not every

amount should be considered as number. Fundamental values up to $3^2 = 9$ are scratches or modes in a conch shell (a 'natural' resonator), and all greater than that would form the conch sell itself. Further evidence should be in the linguistic,
metsys gnitirw a evah dluohs yeht srebmun gnitnuoc fo noitcerid eht ni (tfel ot thgir) in English
starting from a scratch(-) as a unit less than a pair implies an unpaired. Equal, both sides same(=) implies pair, (+) first composite letter of the language,
($\perp$) False. Creators of numbers system from such hypothetical civilization included the understanding into the art, culture and linguistics. Due to the highest variation even if any information is lost after any chaotic event, but the understanding will remain through cultural activities, wired rituals and in language. The main goal is to preserve the understanding in a span of thousands of years. The position of a unit in a number is more important. The understanding of fundamental structure is a recovery from the linguistics rather than anything new.

Today natural variation from the gene pool is removed at a rate of $\ln(n)$, slowly if it reaches at a point of inverse transformation then such process would be irrecoverable.

Here all fields, forces and interactions are shown as unified using the prime function $f(n)$

Chapter 6
Final Value Approximation Method

241 years ago, our beloved Old man Euler rested in a heavenly palace near Saint Petersburg. But let's recall him once again. And try our best to understand more of what he had left behind.

Every value we take at this moment is the initial value, and for every next point, there exists an initial point.

Here we are approximating a plot using the final value approximation method for a specific function of exponent.

$$y'(x) = -2y(x)$$

with the initial condition $y(0) = 1$. The exact solution is:

$$y(x) = e^{-2x}$$

The goal is to compare the numerical approximation using the Modified Euler Method with the exact solution.

6.1 HTML Structure

The code starts by importing the `Plotly.js` library and defining a `div` element where the plot will be displayed:

```html
<!DOCTYPE html>
<html lang="en">
<head>
    <meta charset="UTF-8">
    <meta name="viewport" content="width=device-width,initial-scale=1.0">
    <title>Modified Euler Method Plot</title>
    <script src="https://cdn.plot.ly/plotly-latest.min.js"></script>
</head>
<body>

<div id="plot" style="width: 100%; height: 100%;"></div>
```

6.2 Initialization Using JavaScript

We begin by setting up the initial conditions $x_0 = 0$, $y_0 = 1$, and the number of steps for the Modified Euler Method.

```
// Euler method parameters
let x0 = 0;
```

```
let y0 = 1;
let num_steps = 20;
let a = 0.02; // Guessed neighborhood value

let x_values_modified = [x0];
let y_values_modified = [y0];
```

6.3 Modified Euler Method

Using a loop, we compute the modified Euler method values:

$$y_{n+1} = y_n + h \cdot (-2y_n)$$

with step size h calculated as:

$$h = \frac{x_n - a}{n + 1}$$

```
// Modified Euler's method with variable step
    size
for (let i = 0; i < num_steps; i++) {
    let h = (x_values_modified[x_values_modified
    .length - 1] - a) / (i + 1);
    let y_new = y_values_modified[
    y_values_modified.length - 1] + h * (-2 *
    y_values_modified[y_values_modified.length -
    1]);
    let x_new = x_values_modified[
    x_values_modified.length - 1] + h;
    x_values_modified.push(x_new);
    y_values_modified.push(y_new);
}
```

6.4 Exact Solution

The exact solution is calculated for comparison as $y(x) = e^{-2x}$.

```
// Exact solution
for (let i = 0; i <= 100; i++) {
    let x = i * 2 / 100;
    x_exact.push(x);
    y_exact.push(Math.exp(-2 * x));
}
```

6.5 Plotting with Plotly

Finally, we use Plotly to plot both the modified Euler method values and the exact solution.

```
// Plotting
Plotly.newPlot('plot', [modified_trace,
    exact_trace], layout);
```

This method [5] requires further study and improvement. If you run this in a html file using your browser, a nice graph will get your attention. From this we get to know that,

Every final value is a seed for another initial point.

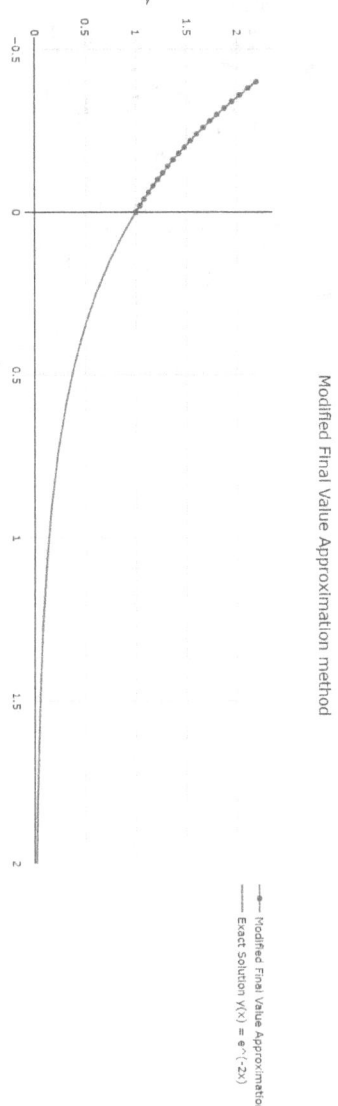

Figure 6.1: The green dots are exact final value approximation of the red curve

Figure 6.2: Leave a review here to provide more understanding regarding these ongoing research

Bibliography

[1] Cern accelerating science.

[2] Sources and types of gravitational waves.

[3] Baryon asymmetry, Jul 2020.

[4] Chronology of the universe, Aug 2020.

[5] Wasiful Alam. A solution for the Initial value problem(https://final-value-approximation.blogspot.com).

[6] Charis Anastopoulos. *Particle or wave: the evolution of the concept of matter in modern physics.* Princeton University Press, 2008.

[7] Kassandra Bell. Magnetic fields may be the key to black hole activity. https://www.nasa.gov/feature/magnetic-fields-may-be-the-key-to-black-hole-activity, 2018.

[8] Niels Bohr. I. on the constitution of atoms and molecules. *The London, Edinburgh, and Dublin Philosophical Magazine and Journal of Science*, 26(151):1–25, 1913.

[9] britanica. radiation pressure, 10 2016.

[10] britanica. Charles's law, 02 2020.

[11] britanica. Lorentz force, 27 2020.

[12] britanica. E=mc2, 7 2021.

[13] britanica. specific heat, 11 2021.

[14] britanica. Specific heat capacity, 11 2021.

[15] I Bernard Cohen and Anne Whitman. The principia. *Newton. A New translation. Proposition 75, Theorem 35: p.956*, 1999.

[16] Edward Salisbury Dana and James Dwight Dana. *A text-book of mineralogy: with an extended treatise on crystallography and physical mineralogy*. J. Wiley & Sons; Chapman & Hall, 1893.

[17] Ashok Das, Thomas Ferbel, and N Gauthier. Introduction to nuclear and particle physics, 1994.

[18] Charles Augustin de Coulomb. Premier mémoire sur l'electricité et le magnétisme. *Histoire de l'Académie Royale des Sciences*, page 569, 1785.

[19] AL Dmitriev. Temperature dependence of gravitational force: experiments, astrophysics, perspectives. *arXiv preprint physics/0611173*, 2006.

[20] Albert Einstein. How i constructed the theory of relativity. *Translated by M. Morikawa. Association of Asia Pacific Physical Societies (AAPPS) Bulletin*, 15(2):17–19, 2005.

[21] Albert Einstein and Edwin P Adams. *Meaning of Relativity (Routledge classics)*. Routledge, 2000.

[22] Eanna E Flanagan and Scott A Hughes. The basics of gravitational wave theory. *New Journal of Physics*, 7(1):204, 2005.

[23] Fredrick Gram. Magnetic fields and forces. *https://web.archive.org/web/20120220030524/ http://instruct.tri- c.edu/fgram/web/Mdipole.htm*, 2012.

[24] Stanley Humphries. Theory and applications of the maxwell stress tensor. *Field Precision, Albuquerque, NM, USA*, 2010.

[25] Paul G Huray. *Maxwell's equations*. John Wiley & Sons, 2011.

[26] VILSON T. ZANCHIN M. MALHEIRO, S. RAY. Electrically charged neutron stars. *https://www.if.ufrgs.br/hadrons/MMalheiro.pdf*.

[27] B Odom, D Hanneke, B d'Urso, and G Gabrielse. New measurement of the electron magnetic moment using a one-electron quantum cyclotron. *Physical Review Letters*, 97(3):030801, 2006.

[28] MIT Opencourseware. Electric charge; electric fields; dipoles; continuous charge distributions. *https://ocw.mit.edu/courses/physics/8-02-physics-ii-electricity-and-magnetism-spring-2007/readings/summary_w01d2.pdf*, 2005.

[29] Wolfgang KH Panofsky and Melba Phillips. *Classical electricity and magnetism*. Courier Corporation, 2005.

[30] R.clarke. The force produced by a magnetic field. *http://info.ee.surrey.ac.uk/Workshop/advice/coils/force.html*, 2010.

[31] Brian Albert Robson. The matter-antimatter asymmetry problem. *Journal of High Energy Physics, Gravitation and Cosmology*, 04(01):166–178, 2018.

[32] NASA Sandra May. What is microgravity. *https://www.nasa.gov/audience/forstudents/5-8/features/nasa-knows/what-is-microgravity-58.html*, 2017.

[33] Eric Sather. The mystery of the matter asymmetry. *Beam Line*, 26:31–37, 1996.

[34] Joseph A Schetz and Allen E Fuhs. *Fundamentals of fluid mechanics*. John Wiley & Sons, 1999.

[35] Emilio Segre. -electron capture by nuclei. *Discovering Alvarez: Selected Works of Luis W.*

Alvarez, with Commentary by His Students and Colleagues, page 11, 1987.

[36] Raymond A Serway and John W Jewett. *Physics for scientists and engineers with modern physics*. Cengage learning, 2018.

[37] J Kenneth Shultis and Richard E Faw. Fundamentals of nuclear science and engineering-kansas state university-manhattan, 2002.

[38] NASA The Imagine Team. Nutron stars. *https://imagine.gsfc.nasa.gov/ science/objects/neutronstars1.html*, 2017.

[39] Patrick Michael Whelan and Michael John Hodgson. *Essential Principles Of Physics*. London, 1987.

[40] Wikipedia. Ferromagnetism. *https://en.wikipedia.org/wiki/Ferromagnetism*, 2018.

[41] wikipedia. Magnetic moment, relation to magnetization. *https://en.wikipedia.org/wiki/Magneticmoment*, 2020.

[42] wikipedia. Paradox of radiation of charged particles in a gravitational field. *https://en.wikipedia.org/wiki/ Paradoxofradiationofchargedparticles inagravitationalfield*, 2020.

[43] Wikipedia. Boltzmann constant, 11 2021.

[44] Wikipedia. Hamiltonian mechanics, 11 2021.

[45] Guan Yiying, Zhang Yang, Li Huawang, Yang Fan, Guan Tianyu, Wang Dongdong, and Teng Hao. Experiment on the relationship between gravity and temperature. *International Journal of Physics*, 6(4):99–104, 2018.

[46] Lawrence Young, Kazuyoshi Yajima, and William Paloski. Artificial gravity research to enable human space exploration. *Paris: International Academy of Astronautics*, pages 1–37, 2009.